Beginnings in Science and its Teaching

Beginnings in Science and its Teaching

Collier Cobb
James J. Walsh

LM Publishers

Some Beginnings in Science[1]

Long before the sciences were pressing their claim to equal rank with ancient learning at Harvard, before Jefferson had seen the establishment of the University of Virginia working under the system of elective studies which he had planned, or before the magnificently endowed institutions of technology were giving what Herbert Spencer regards as knowledge of most worth, we find the beginnings of these things in the newly established university of a State that could boast of only two schools which taught more than the three R's and the very rudiments of the English language.

This modern plan of instruction offered by the University of North Carolina more

---

[1] By Collier Cobb.

than one hundred years ago was the work of a committee of six. Two of this committee were graduates of Princeton, one a graduate and ex-professor of the University of Pennsylvania, two had been students of Harvard, but their education at Cambridge had been interrupted by the Revolutionary War, and the sixth was an eminent lawyer. The names of these men were Samuel McCorkle, David Stone, Alfred Moore, Samuel Ashe, Hugh Williamson, and John Hay.

The course planned by this committee in 1792 gave great prominence to the scientific studies, especially those which could be applied to the arts. The report further recommended the purchase of apparatus for experimental philosophy and astronomy, in which must be included a set of globes, barometer, thermometer, microscope, telescope, quadrant, prismatic glass, electrical machine, and an air-pump. The ancient classics were made elective, the degree of Bachelor of Arts being obtainable without the study of either Latin or Greek. In 1800, however, Latin was made a required study, and an election allowed between French and Greek; and in 1804 Greek was added to the required studies.

It is remarkable that this scheme, adopted in 1792, is almost identical with that adopted by Congress for the colleges to be formed under what is known as the Agricultural and Mechanical College Land Act of 1862. But its interest for us today lies in the fact that it led to the establishment of the first astronomical observatory in the United States, to the first geological survey by public authority in America, and to the first equipment for the teaching of electricity.

JOSEPH CALDWELL.
From the collection of lion.
Kemp   P. Battle.

The men chosen by the trustees to begin this work were David Ker, a graduate of Trinity College, Dublin; Charles W. Harriss, a Princeton man of the class of 1789, Professor of Mathematics; and Samuel A. Holmes, also an alumnus of Princeton. Mr, Harriss was succeeded in his professorship by Joseph Caldwell, Princeton, 1791, who was a tutor at Princeton at the time of his appointment to the professorship in North Carolina.

To Dr. Caldwell we owe the realization of the hopes of the original committee, the ultimate establishment of the observatory, the geological survey, and the electrical laboratory.

A letter written by Prof. Harriss from Chapel Hill, April 10, 1795, shows something of the spirit which Dr. Caldwell was to find in the young university. In it this

Princeton man says: "The constitution of this college is on a more liberal plan than that of any other in America, and by the amendments which I think it will receive at the next meeting of the trustees its usefulness will probably be much promoted. The notion that true learning consists rather in exercising the reasoning faculties and laying up a store of useful knowledge, than in overloading the memory with words of dead languages, is becoming daily more prevalent.

It is hard to deny a young gentleman the honor of a college, after he has with much labor and painful attention acquired a competent knowledge of the sciences, of composing and speaking with propriety in his own language, and has conned the first principles of whatever might render him useful or creditable in the world, merely because he could not read a language two thousand years old."

This letter might well be dated from Boston a century later, for it was nearly a century before such ideas of the essentials of an education were gaining ground with our foremost educators. The literary societies established in 1795 took mottoes in keeping with the spirit of the day, that of the Dialectic Society being "Love of Virtue and Science," and the motto of the Philanthropic Society, "Virtue, Liberty, and Science".

THE OLD TELESCOPES AS THEY ARE TO-DAY IN THE
MITCHELL OBSERVATORY.
Drawn by E. L. Harris.

The first gift to the university, other than lands and money, came from the ladies of Raleigh and Newbern, who contributed a pair of globes, a compass, and a quadrant. The first student, Hinton James, chose as the subjects of his senior forensics, "The Uses of the Sun," "The Commerce of Britain," and "The Motions of the Earth."

ELISHA MITCHELL.
After portrait by Jocelyn.

The young Professor of Mathematics was made president of the university in 1804. His prosperity culminated in 1824, when the financial condition of the university was so good as to allow the trustees to send him to Europe for the purchase of scientific apparatus and books, appropriating six thousand dollars for the purpose.

Soon after his return from Europe President Caldwell planned an observatory, which he built with his own money. The building was finished in 1827, and in the observatory he placed the instruments which he had brought from Europe. These were a meridian transit telescope, made by Simms, of London; an altitude and azimuth telescope, also made by Simms; a telescope for observations on the earth and sky, made by Dolland, of London; and an astronomical clock with a mercurial pendulum, made by

Molineux, of London. To these stationary instruments were added a sextant, made by Wilkinson, of London; a portable reflecting circle, made by Harris, of London; and a Hadley's quadrant.

Before the completion of the observatory building, the clock and meridian transit were set up and used in the library of the university, which was also Prof. Caldwell's lecture room. Here began, in 1825, the first systematic observations upon the heavens made in the United States. Dr. Caldwell was assisted by Profs. Mitchell and Phillips, and their first work was to find the approximate values of the longitude and the latitude of the building in which they worked.

Mitchell was a Yale man of the class of 1813, a native of Connecticut, and a descendant of John Eliot, the apostle to the Indians. Phillips was an Englishman, and a son of a clergyman of the Church of England.

Upon its completion in 1827 the instruments were moved into I he observatory, where observations were made by Dr. Caldwell and his colleagues. The materials used in the building were very poor; the bricks in the wall soon crumbled, and it became necessary, soon after the death of Dr. Caldwell, in January, 1835, to remove the instruments. The building then went rapidly to decay, and fell a victim to fire in 1838.

Observations were, however, continued by Dr. Elisha Mitchell in the attic of the large wooden building which he used as a chemical and metallurgical laboratory. In each end of the attic were two large windows, and in the roof eight others, four on either side. These observations were continued until the summer of 1857, when Prof. Mitchell lost his life upon the highest peak east of the Mississippi River, the mountain which bears his name. By his observations in 1835, 1838, 1844, and 1856 he had established the fact that the peaks of the Black Mountains in North Carolina are the highest east of the Mississippi.

JAMES PHILLIPS.
After portrait by W. G. Brown.

Prof. Phillips has told us that in order "to study the constellations and to show them to his pupils. Dr. Caldwell built on the top of his own residence a platform surrounded by a railing. Here he would sit night after night, pointing out to the seniors, taken in squads of three or four, the outlines of the constellations and their principal stars, and the highway of the planets and the moon. Dr. Caldwell also built in his garden, where they still stand, two pillars of brick, that their eastern and western faces, carefully ground into the same plane, might mark the true meridian.

Near these pillars stood a stone pillar, some five feet high, bearing upon its top a sundial for marking the hours of the day." Before the coming of Mitchell, Princeton thought and Princeton methods had prevailed in the University of North Carolina to the exclusion of all others. But in 1817, Denison Olmsted, a classmate of Mitchell's at Yale, was elected Professor of Chemistry and Geology.

Messrs Mitchell and Olmsted were recommended to Judge William Gaston, then a member of Congress, by the Rev. Sereno Dwight, chaplain of the United States Senate, as young men who were likely to become prominent scientists; and the trustees, upon this recommendation, and upon that of Hon. George E. Badger, who had been their classmate at Yale, offered them chairs in the university.

DENISON OLMSTED.
After a daguerreotype by Moulthrop.

In 1821 Olmsted laid before the Board of Internal Improvements of North Carolina a proposition to undertake a geological and mineralogical survey of the State. This letter is preserved in the executive office at Raleigh. The board approved, and presented the matter to the Legislature. But the

Legislature took no notice of the matter until two years later, when the proposition was renewed.

The survey was authorized by act of the General Assembly, ratified December 31, 1823. Prof. Olmsted was appointed to begin the survey under direction of the State Board of Agriculture, prosecuting the work during the vacations of the university. Thus was established the first geological survey by public authority in America.

It was sustained by an annual appropriation of two hundred and fifty dollars, continued for five years. When Mr. Olmsted resigned in 1825 to accept a professorship at Yale, Dr. Mitchell took up the work of the survey in addition to the duties of his professorship in the university.

Olmsted's report was published in two parts, in 1824 and 1825, and filled in all about one hundred and twenty octavo pages.

FIRST OBSERVATORY.
Drawn from description furnished by John H. Watson,
Esq., Mayor of Chapel Hill, N. C.

The American Journal of Science observes of this survey that, regarded especially as the gratuitous vacation work of a single individual, and in view of the state of geological science in this country at the time, it "must certainly be looked upon as creditable in the highest degree both to the enterprise and to the scientific ability of its projector, and it has undoubtedly been of great benefit not only to the State which authorized it, but to the country and to science generally, by the stimulus which it afforded to similar enterprises in other States."

THE ASTRONOMICAL CLOCK.
This clock still keeps the time for the university.
Drawn by E. L. Harris.

A few years later, in 1829, we find Dr. Caldwell purchasing of W. and S. Jones, mathematical instrument makers, London, the equipment for an electrical laboratory at the University of North Carolina.

The first item on the bill, which lies before me as I write, is "a three-feet plate electrical machine with large branch conductor, supported by two glass pillars, double collectors, mounted in strong mahogany, varnished frame, with six brass legs fitted into brass sockets and screw nuts, negative brass conductor on claw-feet stand from the ground, with connecting sliding tubes, brass bells and wires, etc., £45." The total amount of this first bill for electrical apparatus was £153 4s. 6d.

PROF. MITCHELL'S LABORATORY AND OBSERVATORY.
After photograph by Collier Cobb.

Dr. Caldwell published a Compendious System of Elementary Geometry, in seven books, to which an eighth is added, containing such other propositions as are elementary; subjoined is a Treatise on Plane Trigonometry. He was one of the earliest advocates in the South of popular education by the State.

Dr. Mitchell was the author of a Manual of Chemistry, a second edition of which was passing through the press at the time of his death; a Manual of Geology, illustrated by a geological map of North Carolina; a Manual of Natural History, and a Geography of the Holy Land. Between 1830 and 1840 he contributed many valuable articles to Silliman's Journal.

Denison Olmsted became more widely known than either of the other pioneers in science. In the course of his work at Chapel

Hill he gave the first geological description of the Deep River coal beds, and of the accompanying New Red sandstone, and referred the strata correctly to the same age with the Richmond coal beds and the Connecticut River sandstones. He began researches to determine the practicability of obtaining illuminating gas from cotton seed, but removed to New Haven before he had secured definite results. His Natural Philosophy, which is still a standard work, appeared in 1831, and his Astronomy, another important work, in 1839.

One wonders why such good beginnings should have borne so little fruit; but when we bear in mind that the institution which thus early fostered science had the greater part of its endowment fund swept away by the civil war, that the spirit of the South since that great event has been largely commercial and industrial, and that the income of the old university, from legislative appropriations, tuition fees, and endowment funds, is only forty-five thousand dollars, the wonder ceases.

Science at the Medieval Universities[2]

With the growth of interest in science and in nature study in our own day, one of the expressions that is probably oftenest heard is surprise that the men of preceding generations and especially university men did not occupy themselves more with the world around them and with the phenomena that are so tempting to curiosity.

Science is usually supposed to be comparatively new and nature study only a few generations old. Men are supposed to have been so much interested in book knowledge and in speculations and theories of many kinds, that they neglected the realities of life around them while spinning

---

[2] By James J. Walsh

fine webs of theory. Previous generations, of course, have indulged in theory, but then our own generation is not entirely free from that amusing occupation. Nothing could well be less true, however, that the men of preceding generations were not interested in science even in the sense of physical science, or that nature study is new, or that men were not curious and did not try to find out all they could about the phenomena of the world around them.

The medieval universities and the school-men who taught in them have been particularly blamed for their failure to occupy themselves with realities instead of with speculation. We are coming to recognize their wonderful zeal for education, the large numbers of students they attracted, the enthusiasm of their students since they made so many handwritten copies of the books of their masters, the devotion of the teachers themselves, who wrote at much greater length than do our professors even now and on the most abstruse subjects, so that it is all the more surprising to think they should have neglected science. The thought of our generation in the matter, however, is founded entirely on an assumption. Those who know anything about the writers of the Middle Ages at first hand are not likely to think of them as neglectful of science even in our sense of the term. Those who know them at second hand are, however, very sure in the matter.

The assumption is due to the neglect of science that came in the seventeenth and eighteenth centuries. We have many other similar assumptions because of the neglect of many phases of mental development and applied science at this time. For instance, most of us are very proud of our modern hospital development and think of this as a great humanitarian evolution of applied medical science. We are very likely to think that this is the first time in the world's history that the building of hospitals has been brought to such a climax of development, and that the houses for the ailing in the olden time were mere refuges, prone to become death traps and at most makeshifts for the solution of the problem of the care of the ailing poor. This is true for the hospitals of the seventeenth and eighteenth centuries, but it is not true at all for the hospitals of the thirteenth and fourteenth and fifteenth centuries. Miss Nutting and Miss Dock in their "*History of Nursing*" have called attention to the fact that the lowest period in hospital development is during the eighteenth and early nineteenth centuries. Hospitals were little better than prisons, they had narrow windows, were ill provided with light and air and hygienic arrangements and in general were all that we should imagine old-time hospitals to be. The hospitals of the earlier time, however, had fine high ceilings, large windows, abundant light and air, excellent arrangements for the privacy of patients, and in general were as worthy of the architects of the earlier times as the municipal buildings, the cathedrals, the castles, the university buildings and every other form of construction that the late medieval centuries devoted themselves to.

The trouble with those who assume that there was no study of science and practically no attention to nature study in the Middle Ages is that they know nothing at all about the works of the men who wrote in the medieval period at first hand. They have accepted declarations with regard to the absolute dependence of the scholastics on authority, their almost divine worship of Aristotle, their utter readiness to accept authoritative assertions provided they came with the stamp of a mighty name, and then their complete lack of attention to observation and above all to experiment. Nothing could well be more ridiculous than this ignorant assumption of knowledge with regard to the great teachers at the medieval universities. Just as soon as there is definite knowledge of what these great teachers wrote and taught, not only does the previous mood of blame for them for not paying much more attention to science and nature at once disappear, but it gives place to the heartiest admiration for the work of these great thinkers. It is easy to appreciate then, what Professor Saintsbury said in a recent volume on the thirteenth century:

> And there have even been in these latter days some graceless ones who have asked whether the science of the nineteenth century after an equal interval will be of any more positive value — whether it will not have even less comparative interest than that which appertains to the scholasticism of the thirteenth.

Three men were the great teachers in the medieval universities at their prime. They have been read and studied with interest ever since. They wrote huge tomes, but men have pored over them in every generation. They were Albertus Magnus, the teacher of the other two, Thomas Aquinas and Roger Bacon. All three of them were together at the University of Paris shortly after the middle of the thirteenth century. Anyone who wants to know anything about the attitude of mind of the medieval universities, their professors and students and of all the intellectual world of the time towards science and observation and experiment, should read the books of these men. Any other mode of getting at any knowledge of the real significance of the science of this time is mere pretense. These constitute the documents behind any scientific history of the development of science at this time.

It is extremely interesting to see the attitude of these men with regard to authority. In Albert's tenth book (of his "Summa") in which he catalogues and describes all the trees, plants and herbs known in his time he observes: "All that is here set down is the result of our own experience, or has been borrowed from authors whom we know to have written what their personal experience has confirmed ; for in these matters experience alone can be of certainty."

In his impressive Latin phrase "*experimentum solum certificat in talihiis.*" With regard to the study of nature in general he was quite as emphatic. He was a theologian as well as a scientist, yet in his treatise on "*The Heavens and The Earth*" he declared that "in studying nature we have not to inquire how God the Creator may, as He freely wills, use His creatures to work miracles and thereby show forth His power. We have rather to inquire what nature with its immanent causes can naturally bring to pass."

Just as striking quotations on this subject might be made from Roger Bacon. Indeed, Bacon was quite impatient with the scholars around him who talked over much, did not observe enough, depended to excess on authority and in general did as mediocre scholars always do, made much fuss on second-hand information — plus some filmy speculations of their own. Friar Bacon, however, had one great pupil whose work he thoroughly appreciated because it exhibited the opposite qualities.

This was Petrus — we have come to know him as Peregrinus — whose observations on magnetism have excited so much attention in recent years with the republications of his epistle on the subject. It is really a monograph on magnetism written in the thirteenth century.

Roger Bacon's opinion of it and of its author furnishes us the best possible index of his attitude of mind towards observation and experiment in science.

I know of only one person who deserves praise for his work in experimental philosophy for he does not care for the discourses of men and their wordy warfare, but quietly and diligently pursues the works of wisdom. Therefore what others grope after blindly, as bats in the evening twilight, this man contemplates in all their brilliancy because he is a master of experiment. Hence, he knows all of natural science whether pertaining to medicine and alchemy, or to matters celestial or terrestrial. He has worked diligently in the smelting of ores as also in the working of minerals; he is thoroughly acquainted with all sorts of arms and implements used in military service and in hunting, besides which he is skilled in agriculture and in the measurement of lands.

It is impossible to write a useful or correct treatise in experimental philosophy without mentioning this man's name.

Moreover, he pursues knowledge for its own sake; for if he wished to obtain royal favor, he could easily find sovereigns who would honor and enrich him.

Similar expressions might readily be quoted from Thomas Aquinas, but his works are so easy to secure and his whole attitude of mind so well known, that it scarcely seems worth while taking space to do so. Aquinas is still studied very faithfully in many universities and within the last few years one of his great text-books of philosophy has been replaced in the curriculum of Oxford University, in which it occupied a prominent position in the long ago, as a work that may be offered for examination in the department of philosophy. It is with regard to him particularly that there has been the greatest revulsion of feeling in recent years and a recognition of the fact that here was a great thinker familiar with all that was known in the physical sciences, and who had this knowledge constantly in his mind when he drew his conclusions with regard to philosophical and theological questions.

As for the supposed swearing by Aristotle which is so constantly asserted to have been the habit of these scholastic philosophers, it is extremely difficult in the light of expressions which they themselves use to understand how this false impression arose, Aristotle they thoroughly respected. They constantly referred to his works, but so has every thinking generation ever since, Whenever he had made a declaration they would not accept the contradiction of it without a good reason, but whenever they had good reasons, Aristotle's opinion was at once rejected without compunction, Albertus Magnus, for instance, said : "Whoever believes that Aristotle was a God must also believe that he never erred, but if we believe that Aristotle was a man, then doubtless he was liable to err just as we are."

A number of direct contradictions of Aristotle we have from Albert. A well-known one is that with regard to Aristotle's assertion that lunar rainbows appeared only twice in fifty years, Albert declared that he himself had seen two in a single year.

Indeed, it seems very clear that the whole trend of thought among the great teachers of the time was away from acceptance of scientific conclusions on authority unless there was good evidence for them available. They were quite as impatient as the scientists of our time with a constant putting forward of Aristotle as if that settled the scientific question.

Roger Bacon wanted the Pope to forbid the study of Aristotle because his works were leading men astray from the study of science, his authority being looked upon as so great that men did not think for themselves but accepted his assertions. Smaller men are always prone to do this, and indeed it constitutes one of the difficulties in the way of advance in scientific knowledge, as Roger Bacon himself pointed out.

These are the sort of expressions that are to be expected from Friar Bacon from what we know of other parts of his work. His "*Opus Tertium*" was written at the request of Pope Clement IV, because the Pope had heard many interesting accounts of what the great thirteenth-century teacher and experimenter was doing at the University of Oxford, and wished to learn for himself the details of his work.

Bacon starts out with the principle that there are four grounds of human ignorance. These are, "first, trust in inadequate authority; second, that force of custom which leads men to accept without properly questioning what has been accepted before their time ; third, the placing of confidence in the assertions of the inexperienced; and fourth, the hiding of one's own ignorance behind the parade of superficial knowledge, so that we are afraid to say I do not know." Professor Henry Morley, a careful student of Bacon's writings, said with regard to these expressions of Bacon :

> No part of that ground has yet been cut away from beneath the feet of students, although six centuries have passed. We still make sheep-walks of second, third and fourth, and fiftieth hand references to authority; still we are the slaves of habit, still we are found following too frequently the untaught crowd, still we flinch from the righteous and wholesome phrase " I do not know " and acquiesce actively in the opinion of others that we know what we appear to know.

In his *"Opus Majus"* Bacon had previously given abundant evidence of his respect for the experimental method. There is a section of this work which bears the title *Scientia Experimentalis*. In this Bacon affirms that "without experiment nothing can be adequately known. An argument may prove the correctness of a theory, but does not give the certitude necessary to remove all doubt, nor will the mind repose in the clear view of truth unless it finds its way by means of experiment." To this he later added in his "*Opus Tertium*": "The strongest argument proves nothing so long as the conclusions are not verified by experience. Experimental science is the queen of sciences, and the goal of all speculation."

It is no wonder that Dr. Whewell in his "*History of the Inductive Sciences*" should have been unstinted in his praise of Roger Bacon's work and writings. In a well-known passage he says of the "Opus Majus":

Roger Bacon's "Opus Majus" is the encyclopedia and "Novum Organon" of the thirteenth century, a work equally wonderful with regard to its wonderful scheme and to the special treatises by which the outlines of the plans are filled up. The professed object of the work is to urge the necessity of a reform in the mode of philosophizing, to set forth the reasons why knowledge had not made greater progress, to draw back attention to the sources of knowledge which had been unwisely neglected, to discover other sources which were yet almost untouched, and to animate men in the undertaking of a prospect of the vast advantages which it offered. In the development of this plan all the leading portions of science are expanded in the most complete shape which they had at that time assumed; and improvements of a very wide and striking kind are proposed in some of the principal branches of study. Even if the work had no leading purposes it would have been highly valuable as a treasure of the most solid knowledge and soundest speculations of the time; even if it had contained no such details it would have been a work most remarkable for its general views and scope.

As a matter of fact the universities of the middle ages, far from neglecting science, were really scientific universities. Because the universities of the early nineteenth century occupied themselves almost exclusively with languages and especially formed students' minds by means of classical studies, we in our generation are prone to think that such linguistic studies formed the main portion of the curriculum of the universities in all the old times and particularly in the middle ages. The study of the classic languages, however, came into university life only after the renaissance. Before that the undergraduates of the universities had occupied themselves almost entirely with science.

It was quite as much trouble to introduce linguistic studies into the old universities in the renaissance time to replace science, as it was to secure room for science by pushing out the classics in the modem time. Indeed the two revolutions in education are strikingly similar when studied in detail. Men who had been brought up on science before the renaissance were quiet sure that that formed the best possible means of developing the mind. In the early nineteenth century men who had been formed on the classics were quite as sure that science could not replace them with any success.

There is no pretense that this view of the medieval universities is a new idea in the history of education. Those who have known the old universities at first hand by the study of the actual books of their professors and by familiarity with their courses of study, have not been inclined to make the mistake of thinking that the medieval university neglected science. Professor Huxley in his "*Inaugural Address as Rector of Aberdeen University*" some thirty years ago stated very definitely his recognition of medieval devotion to science. His words are well worth remembering by all those who are accustomed to think of our time as the first in which the study of science was taken up seriously in our universities. Professor Huxley said :

> The scholars of the medieval universities seem to have studied grammar, logic and rhetoric; arithmetic and geometry; astronomy, theology and music. Thus their work, however imperfect and faulty, judged by modem lights, it may have been, brought them face to face with all the leading aspects of the many-sided mind of man. For these studies did really contain, at any rate in

embryo, sometimes it may be in caricature, what we now call philosophy, mathematical and physical science and art. *And I doubt if the curriculum of any modem university shows so clear and generous a comprehension of what is meant by culture, as this old Trivium and Quadrivium does.*

It would be entirely a mistake, however, to think that these great writers and teachers who influenced the medieval universities so deeply and whose works were the text-books of the universities for centuries after, only had the principles of physical and experimental science and did not practically apply them. As a matter of fact their works are full of observation. Once more, the presumption that they wrote only nonsense with regard to science comes from those who do not know their writings at all, while great scientists who have taken the pains to study their works are enthusiastic in praise. Humboldt, for instance, says of Albertus Magnus, after reading some of his works with care:

Albertus Magnus is equally active and influential in promoting the study of natural science and of the Aristotelian philosophy. His works contain some exceedingly acute remarks on the organic structure and physiology of plants. One of his works bearing the title of "Liber Cosmographicus De Natura Locorum" is a species of physical geography. I have found in it considerations on the dependence of temperature concurrently on latitude and elevation and on the effect of different angles of the sun's rays in heating the ground which have excited my surprise.

It is with regard to physical geography of course that Humboldt is himself a distinguished authority.

Humboldt's expression that he found some exceedingly acute re- marks on the organic structure and physiology of plants in Albert the Great's writings will prove a great surprise to many people.

Meyer, the German historian of botany, however, has reechoed Humboldt's praise with emphasis. The extraordinary erudition and originality of Alberts treatise on plants drew from Meyer the comment :

No botanist who lived before Albert can be compared with him unless Theophrastus, with whom he was not acquainted; and after him none has painted nature in such living colors or studied it so profoundly until the time of Conrad Gessner and Csesalpino.

These men, it may be remarked, come three centuries after Albert's time. A ready idea of Albert's contributions to physical science can be obtained from his life by Sighart, which has been translated into English by Dixon and was published in London in 1870.

Pagel, in Puschmann's "History of Medicine, *already referred to, gives a list* of the books written by Albert on scientific matters with some comments which are eminently suggestive, and furnish solid basis for the remark that I have made, that men's minds were occupied with nearly the same problems in science in the thirteenth century as we are now, while the conclusions they came to were not very different from ours, though reached so long before us.

This catalogue of Albertus Magnus's works shows very well his own interest and that of his generation in physical science of all kinds. There were eight treatises on Aristotle's physics and on the underlying principles of natural philosophy and of energy and of movement; four treatises concerning the heavens and the earth, one on physical geography which also contains, according to Pagel, numerous suggestions on ethnography and physiology. There are two treatises on generation and corruption, six books on meteors, five books on minerals, three books on the soul, two books on the intellect, a treatise on nutritives, and then a treatise on the senses and another on the memory and on the imagination.

All the phases of the biological sciences were especially favorite subjects of his study. There is a treatise on the motion of animals, a treatise in six books on vegetables and plants, a treatise on breathing things, a treatise on sleep and waking, a treatise on youth and old age and a treatise on life and death. His treatise on minerals contains, according to Pagel, a description of ninety-five different kinds of precious stones.

Albert's volumes on plants were reproduced with Meyer, the German botanist, as editor (Berlin, 1867). All of Albert's books are available in modern editions.

Pagel says of Albertus Magnus that: His profound scholarship, his boundless industry, the almost incontrollable impulse of his mind after universality of knowledge, the many-sidedness of his literary productivity, and finally the almost universal recognition which he received from his contemporaries and succeeding generations, stamp him as one of the most imposing characters and one of the most wonderful phenomena of the middle ages.

In another passage Pagel has said :

While Albert was a Churchman and an ardent devotee of Aristotle, in matters of natural phenomena he was relatively unprejudiced and presented an open mind. He thought that he must follow Hippocrates and Galen, rather than Aristotle and Augustine, in medicine and in the natural sciences. We must concede it a special subject of praise for Albert that he distinguished very strictly between natural and supernatural phenomena. The former he considered as entirely the object of the investigation of nature. The latter he handed over to the realm of metaphysics.

Roger Bacon is, however, the one of these three great teachers who shows us how thoroughly practical was the scientific knowledge of the universities and how much it led to important useful discoveries in applied science and to anticipations of what is most novel even in our present-day sciences. Some of these indeed are so startling, that only that we know them not by tradition but from his works, where they may be readily found without any doubt of their authenticity, we should be sure to think that they must be the result of later commentators' ideas.

Bacon was very much interested in astronomy, and not only suggested the correction of the calendar, but also a method by which it could be kept from wandering away from the actual date thereafter. He discovered many of the properties of lenses and is said to have invented spectacles and announced very emphatically that light did not travel instantaneously but moved with a definite velocity. He is sometimes said to have invented gunpowder, but of course he did not, though he studied this substance in various forms very carefully and drew a number of conclusions in his observations. He was sure that some time or other man would learn to control the energies exhibited by explosives and that then he would be able to accomplish many things that seemed quite impossible under their present conditions. He said, for instance :

Art can construct instruments of navigation, such that the largest vessela governed by a single man will traverse rivers and seas more rapidly than if they were filled with oarsmen. One may also make carriages which without the aid of any animal will run with remarkable swiftness.

In these days when the automobile is with us and when the principal source of energy for motor purposes is derived from explosives of various kinds this expression of Roger Bacon represents a prophecy marvelously surprising in its fulfillment. It is no wonder that the book whence it comes bears the title "*De Secretis Artis et Xaturae.*"

Roger Bacon even went to the extent, however, of declaring that man would some time be able to fly. He was even sure that with sufficient pains he could himself construct a flying machine. He did not expect to use explosives for his motor power, however, but thought that a windlass properly arranged, worked by hand, might enable a man to make sufficient movement to carry himself aloft or at least to sup- port himself in the air, if there were enough surface to enable him to use his lifting power to advantage. He was in intimate relations by letter with many other distinguished inventors and investigators besides Peregrinus and was a source of incentive and encouragement to them all.

The more one knows of Aquinas the more surprise there is at his anticipation of many modem scientific ideas. At the conclusion of a course on cosmology delivered at the University of Paris he said that "nothing at all would ever be reduced to nothingness" (nihil omnino in nihilum redigeiur). He was teaching the doctrine that man could not destroy matter and God would not annihilate it. In other words, he was teaching the indestructibility of matter even more emphatically than we do. He saw the many changes that take place in material substances around us, but he taught that these were only changes of form and not substantial changes and that the same amount of matter always remained in the world. At the same time he was teaching that the forms in matter by which he meant the combinations of energies which distinguish the various kinds of matter are not destroyed. In other words, he was anticipating not vaguely, but very clearly and definitely, the conservation of energy. His teaching with regard to the composition of matter was very like that now held by physicists. He declared that matter was composed of two principles, prime matter and form. By forma he meant the dynamic element in matter, while by materia prima he meant the underlying substratum of material, the same in every substance, but differentiated by the dynamics of matter.

It used to be the custom to make fun of these medieval scientists for believing in the transmutation of metals. It may be said that all three of these greatest teachers did not hold the doctrine of the trans- mutation of metals in the exaggerated way in which it appealed to many of their contemporaries. The theory of matter and form, how- ever, gave a philosophical basis for the idea that one kind of matter might be changed into another.

We no longer think that notion absurd. Sir "William Ramsay has actually succeeded in changing one element into another and radium and helium are seen changing into each other, until now we are quite ready to think of transmutation placidly. The Philosopher's Stone used to seem a great absurdity until our recent experience with radium, which is to some extent at least the philosopher's stone, since it brings about the change of certain supposed elements into others. A distinguished American chemist said not long ago that he would like to extract all the silver from a large body of lead ore in which it occurs so commonly, and then come back after twenty years and look for further traces of silver, for he felt sure that they would be found and that lead ore is probably always producing silver in small quantities and copper ore is producing gold.

Most people will be inclined to ask where the fruits of this under- graduate teaching of science are to be found. They are inclined to presume that science was a closed book to the men and women of that time. It is not hard, however, to point the effect of the scientific training in the writings of the times. Dante is a typical university man of the period. He was at several Italian universities, was at Paris and perhaps at Oxford. His writings are full of science. Professor Kühns, of Wesleyan, in his book "The Treatment of Nature in Dante," has pointed out how much Dante knows of science and of nature. Few of the poets not only of his own but of any time have known more. There are only one or two writers of poetry in our time who go with so much confidence to nature and the scientific interpretation of her for figures for their poetry.

The astronomy, the botany, the zoology of Albertus Magnus and Thomas Aquinas, Dante knew very well and used confidently for figurative purposes. Anyone who is inclined to think nature study a new idea in the world forgets, or has never known, his Dante. The birds and the bees, the flowers, the leaves, the varied aspects of clouds and sea, the phenomena of phosphorescence, the intimate habits of bird and beast and the ways of the plants, as well as all the appearances of the heavens, Dante knew very well and in a detail that is quite surprising when we recall how little nature study is supposed to have attracted the men of his time. Only that his readers appreciated it all, Dante would surely not have used his scientific erudition so constantly.

So much for the undergraduate department of the universities of the middle ages, and the view is absolutely fair, for these were the men to whom the students flocked by thousands. They were teaching science, not literature. They were discussing physics as well as metaphysics, psychology in its phenomena as well as philosophy, observation and experiment as well as logic, the ethical sciences, economics, practically all the scientific ideas that were needed in their generation — and that generation saw the rise of the universities, the finishing of the cathedrals, the building of magnificent town halls and castles and beautiful municipal buildings of many kinds, including hospitals, the development of the Hansa League in commerce and of wonderful manufacturers of all the textiles, the arts and crafts, as well as the most beautiful book-making and art and literature. We could be quite sure that the men who solved all the other problems so well could not have been absurd only in their treatment of science.

Anyone who reads their books will be quite sure of that.

While most people might be ready then to confess that possibly Huxley was not mistaken with regard to the undergraduate department of the universities, most of them would feel sure that at least the graduate departments were sadly deficient in accomplishment. Once more this is entirely an assumption. The facts are all against any such idea.

There were three graduate departments in most of the universities — theology, law and medicine. While physical scientists are usually not cognizant of it apparently, theology is a science, a department of knowledge developed scientifically, and most of these medieval universities did more for its scientific development than the schools of any other period. Quite as much may be said for philosophy, for there are many who hesitate to attribute any scientific quality to modem developments in the matter. As for law, this is the great period of the foundation of scientific law development, the English common law was formulated by Bracton, the deep foundations of basic French and Spanish law were laid, and canon law acquired a definite scientific character which it was always to retain. All this was accomplished almost entirely by the professors in the law departments of the universities.

It was in medicine, however, where most people would be quite sure without any more ado that nothing worth while talking about was being done, that the great triumphs of graduate teaching at the medieval universities were secured. Here more than anywhere else is there room for supreme surprise at the quite unheard-of anticipations of our modern medicine and stranger still, as it may seem, of our modern surgery.

The law regulating the practise of medicine in the Two Sicilies about the middle of the thirteenth century shows us the high standard of medical education. Students were required to have three years of preliminary study at the university, four years in the medical department and then practise for a year with a physician before they were allowed to practise for themselves. If they wanted to practise surgery, an extra year in the study of anatomy was required.

I published the text of this law, which was issued by the Emperor Frederick II about 1241, in the Journal of the American Medical Association three years ago. It also regulated the practise of pharmacy. Drugs were manufactured under the inspection of the government and there was a heavy penalty for substitution, or for the sale of old inert drugs, or improperly prepared pharmaceutical materials. If the government inspector violated his obligations as to the oversight of drug preparations the penalty was death. Nor was this law of the Emperor Frederick an exception. We have the charters of a number of medical schools issued by the popes during the next century, all of which require seven years or more of university study, four of them in the medical department before the doctor's degree could be obtained.

When new medical schools were founded they had to have professors from certain well- recognized schools on their staff at the beginning in order to assure proper standards of teaching, and all examinations were conducted under oath-bound secrecy and with the heaviest obligations on professors to be assured of the knowledge of students before allowing them to pass.

It might be easy to think, and many people are prone to do so, that in spite of the long years of study required there was really very little to study in medicine at that time. Those who think so should read Professor Clifford Allbutt's address on the "*Historical Relations of Medicine and Surgery*" delivered at the World's Fair at St. Louis in 1904. He has dwelt more on surgery than on medicine, but he makes it very clear that he considers that the thinking professors of medicine of the later Middle Ages were doing quite as serious work in their way as any that has been done since. They were carefully studying cases and writing case histories, they were teaching at the bedside, they were making valuable observations and they were using the means at their command to the best advantage. Of course there are many absurdities in their therapeutics, but then we must not forget there have always been many absurdities in therapeutics and that we are not free from them in our day. Professor Richet, at the University of Paris, said not long ago "the therapeutics of any generation is quite absurd to the second succeeding generation." We shall not blame the medieval generations for having accepted remedies that afterwards proved inert, for every generation has done that, even our own.

Their study of medicine was not without lasting accomplishment however. They laid down the indications and the dosage for opium. They used iron with success, they tried out many of the bitter tonics among the herbal medicines, and they used laxatives and purgatives to good advantage. Down at Montpelier, Gilbert, the Englishman, suggested red light for smallpox because it shortened the fever, lessened the lesions and made the disfigurement much less. Finsen was given the Nobel prize partly for rediscovery of this. They segregated erysipelas and so prevented its spread. They recognized the contagiousness of leprosy and though it was probably as wide-spread as tuberculosis is at the present time, they succeeded not only in controlling but in eventually obliterating it throughout Europe.

It was in surgery, however, that the greatest triumphs of teaching of the medieval universities were secured. Most people are inclined to think that surgery developed only in our day. The great surgeons of the thirteenth and fourteenth century, however, anticipated most of our teaching. They investigated the causes of the failure of healing by first intention, recognized the danger of wounds of the neck, differentiated the venereal diseases, described rabies and knew much of blood poisoning, and operated very skillfully.

We have their text-books of surgery and they are a never-ending source of surprise. They operated on the brain, on the thorax, on the abdominal cavity, and did not hesitate to do most of the operations that modem surgeons do. They operated for hernia by the radical cure, though Mondeville suggested that more people were operated on for hernia for the benefit of the doctor's pocket than for the benefit of the patient. Guy de Chauliac declared that in wounds of the intestines patients would die unless the intestinal lacerations were sewed up and he described the method of suture and invented a needle holder. We have many wonderful instruments from these early days preserved in pictures at least, that show us how much modern advance is merely reinvention.

They understood the principles of aseptic surgery very well. They declared that it was not necessary "that pus should be generated in wounds." Professor Clifford Allbutt says:

They washed the wound with wine, scrupulously removing every foreign particle; then they brought the edges together, not allowing wine or anything else to remain within — dry adhesive surfaces were their desire. Nature, they said, produces the means of union in a viscous exudation, or natural balm, as it was afterwards called by Paracelsus, Paré and Wurtz. In older wounds they did their best to obtain union by cleansing, desiccation and refreshing of the edges. Upon the outer surface they laid only lint steeped in wine. Powders they regarded as too desiccating, for powder shuts in decomposing matters; wine after washing purifying and drying the raw surfaces evaporates.

Almost needless to say these are exactly the principles of aseptic surgery. The wine was the best antiseptic that they could use and we still use alcohol in certain cases. It would seem to many quite impossible that such operations as are described could have been done without anesthetics, but they were not done without anesthetics. There were two or three different forms of anesthesia used during the thirteenth and fourteenth centuries. One method employed by Ugo da Lucca consisted of the use of an inhalant. We do not know what the material employed was. There are definite records, however, of its rather frequent employment.

What a different picture of science at the medieval universities all this makes from what we have been accustomed to hear and read with regard to them. It is difficult to understand where the old false impressions came from. The picture of university work that recent historical research has given us shows us professors and students busy with science in every department, making magnificent advances, many of which were afterwards forgotten, or at least allowed to lapse into desuetude.

The positive assertions with regard to old-time ignorance were all made in the course of religious controversy. In English-speaking countries particularly it became a definite purpose to represent the old church as very much opposed to education of all kinds and above all to scientific education. There is not a trace of that to be found anywhere, but there were many documents that were appealed to to confirm the protestant view. There was a papal bull, for instance, said to forbid dissection. When read it proves to forbid the cutting up of bodies to carry them to a distance for burial, an abuse which caused the spread of disease, and was properly prohibited. The church prohibition was international and therefore effective. At the time the bull was issued there were twenty medical schools doing dissection in Italy and they continued to practise it quite undisturbed during succeeding centuries. The papal physicians were among the greatest dissectors. Dissections were done at Rome and the cardinals attended them. Bologna at the height of its fame was in the Papal States.

All this has been ignored and the supposed bull against anatomy emphasized as representing the keynote of medical and surgical history. Then there was a papal decree for- bidding the making of gold and silver. This was said to forbid chemistry or alchemy and so prevent scientific progress. The history of the medical schools of the time shows that it did no such thing. The great alchemists of the time doing really scientific work were all clergymen, many of them very prominent ecclesiastics.

Just in the same way there were said to be decrees of the church councils forbidding the practise of surgery. President White says in his "Warfare of Religion with Theology in Christendom" that as a consequence of these surgery was in dishonor until the Emperor "Winceslaus, at the beginning of the fifteenth century, ordered that it should be restored to estimation. As a matter of fact during the two centuries immediately preceding the first years of the fifteenth century, surgery developed very wonderfully and we have probably the most successful period in all the history of surgery except possibly our own. The decrees forbade monks to practise surgery because it led to certain abuses. Those who found these decrees and wanted to believe that they prevented all surgical development simply quoted them and assumed there was no surgery. The history of surgery at this time is one of the most wonderful chapters in human progress.

The more we know of the Middle Ages the more do we realize how much they accomplished in every department of intellectual effort. Their development of the arts and crafts has never been equalled in the modern time. They made very great literature, marvelous architecture, sculpture that rivals the Greeks, painting that is still the model for our artists, surpassing illuminations; everything that they touched became so beautiful as to be a model for all the after time. They accomplished as much in education as they did in all the other arts, their universities had more students than any that have existed down to our own time and they were enthusiastic students and their professors were ardent teachers, writers, observers, investigators. While we have been accustomed to think of them as neglecting science their minds were occupied entirely with science. They succeeded in anticipating much more of our modern thought and even scientific progress than we have had any idea until comparatively recent years. The work of the later middle ages in mathematics is particularly strong and was the incentive for many succeeding generations. Roger Bacon insisted that without mathematics there was no possibility of real advance in physical science. They had the right ideas in every way. While they were occupied more with the philosophical and ethical sciences than we are, these were never pursued to the neglect of the physical sciences in the strictest sense of that term.

Is it not time that we should drop the foolish notions that are very commonly held because we know nothing about the middle ages — and therefore the more easily assume great knowledge, and get back to appreciate the really marvellous details of educational and scientific development which are so interesting and of so much significance at this time?

THE JOB I CALLED A

CALLED A

DREAM

Removing Shiny Object
Syndrome for Entrepreneurs

RYAN HODGSON

RYAN HODGSON

ISBN: 978-1-5331-3852-1